COMPUTER CHIPS

and Other Hardware Tech

Co-published by agreement between Shi Tu Hui and World Book, Inc.

Shi Tu Hui
Room 1807, Block 1,
#3 West Dawang Road
Chaoyang District, Beijing 100025
P.R. China

World Book, Inc
180 North LaSalle Street
Suite 900
Chicago, Illinois 60601
USA

Library of Congress Cataloging-in-Publication Data for this volume has been applied for.

Cool Tech (set #2)
ISBN: 978-0-7166-5387-5 (set, hc)

Computer Chips and Other Hardware Tech
ISBN: 978-0-7166-5390-5 (hc)

Also available as:
ISBN: 978-0-7166-5396-7 (e-book)
ISBN: 978-0-7166-5402-5 (soft cover)

Written by William D. Adams

STAFF

VP, Editorial: Tom Evans

Manager, New Product: Nicholas Kilzer

Curriculum Designer: Caroline Davidson

Proofreader: Nathalie Strassheim

Indexer: Nathaniel Lindstrom

Coordinator, Design Development & Production: Brenda Tropinski

Senior Media Editor: Rosalia Bledsoe

Developed with World Book by
White-Thomson Publishing LTD

www.wtpub.co.uk

ACKNOWLEDGMENTS

Cover   © Pedal to the Stock/Shutterstock

5   German Federal Archives (licensed under CC BY SA 3.0)

6-9   © Shutterstock

10-11   U.S. Army; © Chemical Heritage Foundation Collections; © AP Photo; © Intel Free Press; © Intel

12-13   © Pastry Shop/Shutterstock; © MS Mikel/Shutterstock; © Intel

14-15   © Elena11/Shutterstock; © Intel

16-17   © Stanford Engineering; © Marco-Alexis Chaira, ASU; © Boykov/Shutterstock

18-19   © Piotr Tomicki, Shutterstock; © Anurak Pongpatimet, Shutterstock

20-21   © Simon Mayer, Shutterstock; © Vanessa Volk, Shutterstock; © Claudio Divizia, Shutterstock; © Everett Collection/Shutterstock; United States Census Bureau

22-23   © Claudio Divizia, Shutterstock; Marcin Wichary (licensed under CC BY 2.0); © World History Archive/Alamy Images; © Original Negative/Alamy Images

24-25   © Gorodenkoff/Shutterstock; © Microsoft; Association of the Internet Industry & NeRZ

26-27   © ShotPrime Studio/Shutterstock; © Ruslan Lytvyn, Shutterstock; © Proxima Studio/Shutterstock; © SARANS/Shutterstock; © 5D Memory Crystal

28-29   © Metamorworks/Shutterstock; © Rami Halim, iStockphoto

30-31   © Naruedom Yaempongsa, Shutterstock; © Mark Richards, Zuma Press/Alamy Images; © SRI International and the Doug Englebart Institute; © Tatiana Popova, Shutterstock; © Helder Almeida, Shutterstock; © ZZ, Inc.

32-33   © Ground Picture/Shutterstock; © Perfect Lab/Shutterstock; © MRATHS; Public Domain; © Zuma Press/Alamy Images; © Kaspars Grinvalds, Shutterstock

34-35   © DC Studio/Shutterstock; © Stephen Dorey ABIPP/Alamy Images

36-37   © Alfonso d'Agostino, Shutterstock; © jultud/Shutterstock; © FUJITSU; © Framesira/Shutterstock

38-39   © Gilmanshin/Shutterstock; National Museum of American History; © Xerox; © INTERFOTO/Alamy Images

40-41   © Grzegorz Czapski, Alamy Images; © Samsung; © Travelstock44/Alamy Images; © Patryk Kosmider, Shutterstock

42-43   NASA/Goddard/Bill Hrybyk; Oleg Alexandrov (licensed under CC BY SA 3.0)

44-45   Michael Hicks (licensed under CC BY 2.0); © Riken; National Weather Service; © PBH Images/Alamy Images; © Intel

CONTENTS

Acknowledgments.....................................2

Glossary 4

Introduction..5

(1) Computer Chips 6

(2) Hard Drives 18

(3) Touchscreens and Other Input Devices 28

(4) Monitors and Televisions 34

(5) Supercomputers................................ 42

Engage Your Reader 46

Index... 48

There is a glossary of terms on the first page. Terms defined in the glossary are in boldface type that **looks like this** on their first appearance on any spread (two facing pages).

GLOSSARY

architect a person who designs and lays out plans for buildings.

augmented reality (AR) the addition of artificial visual, auditory, or other sensory information to the physical world, so that it appears to be part of the actual environment.

cathode-ray tube (CRT) a vacuum tube in which high-speed electrons are produced and passed through electromagnetic fields. Cathode-ray tubes were once widely used in television screens and computer monitors.

cloud a network of servers used to store and process information sent over the internet.

data center a facility that houses computer systems and communications equipment, often providing remote or "cloud" computing services.

DNA deoxyribonucleic acid, a thin, chainlike molecule found in every living cell. It directs the formation, growth, and reproduction of cells and organisms.

engineer a professional who plans and builds engines, machines, roads, or the like.

gene a part of a chromosome that influences inheritance and development of some characteristics.

laser a device that produces a very narrow and intense beam of a very narrow range of light wavelengths going in only one direction. By contrast, a standard light source produces light of many wavelengths all traveling in slightly different directions.

plasma in physics, is a form of matter composed of electrically charged particles. The sun and the other stars consist of plasma.

resolution the ability of a lens or sensor to produce separate images of objects that are very close together.

smartphone a portable telephone equipped to perform additional functions beyond calling, such as providing internet access, supporting text messaging, or taking photographs.

transistor a tiny device that controls the flow of electric current in electronic equipment.

virtual reality an artificial, three-dimensional computer environment.

INTRODUCTION

Modern computers have only been around for about 50 years. But in that time, technology has changed in almost unimaginable ways.

Picture yourself as a young computer programmer at the beginning of the computer age. You arrive at work early to finish a program you're writing. Instead of working on a computer, you enter the code by punching tiny holes into dozens of small pieces of cardstock. You work through lunch to beat the afternoon rush at the *mainframe*—a large, central computer. But in your hurry, you drop your deck of punched cards!

By the time you've put the cards back in order, there's a long wait at the mainframe. When your turn arrives, you hand your stack of cards to the operator. You must wait a few minutes while the mainframe calculates your results—and that's if you haven't made any mistakes!

On the way back to your desk, you jealously eye the more experienced programmers. They don't need to use cards at all—instead, they share several computer terminals that connect directly to a newer mainframe. Their terminals have black-and-green monitors and keyboards. You get a small taste of this experience as you write up your report on an electric word processor. It doesn't have a screen, but it's a step up from the typewriter you were using a couple of years ago.

This scenario may seem strange, but people used to work this way not so long ago. In another 50 years, who knows how much computer hardware will change?

1 COMPUTER CHIPS

THE MAGIC INSIDE THE MACHINE

The computer chip is the beating heart of any computer. Computer chips take in information and instructions, perform calculations, and send the results to other parts of a machine.

Chips aren't just found in computers anymore, either. A growing number of devices depend on computer chips to run smartly and efficiently. Computers keep our cars running, monitor the temperature in our homes, and keep our appliances functioning properly.

Not only has society come to rely on computer chips, but we've come to rely on chips getting ever more powerful. Programmers continue to design apps and new features that use more computing power. Remember a computer or **smartphone** from 10 years ago? It probably can't keep up with modern programs.

Many scholars have begun to wonder if advances in computer chips can continue to keep pace with demand. Inventors are working on the next generation of chip technology to address this issue.

CHIP BASICS

Each computer chip contains billions of tiny devices called **transistors.** Transistors control the flow of electric current, either switching the current on and off or *amplifying* (strengthening) the current. A small voltage called the *input signal* controls both switching and amplification.

Transistors at work. Transistors in computer chips perform rapid switching operations to manipulate electric charges. The charges represent information in the form of the 0's and 1's of the binary number system. Everything that passes through the computer chip, from low-level computer languages to computer games, is first translated into binary code. As the transistors move the charges around, electronic circuits carry out calculations, solve problems in logic, form images on screens, and perform other computer operations.

Finding the right material. Computer chips rely on the precise flow of small amounts of electricity to perform calculations. Electricity flows easily through some materials, called *conductors*. But conductors would allow too much electricity to pass, flooding or even damaging the transistors. Other materials resist the flow of electricity. Such materials are called *insulators*. But chips made of insulators couldn't conduct any electricity or perform any calculations.

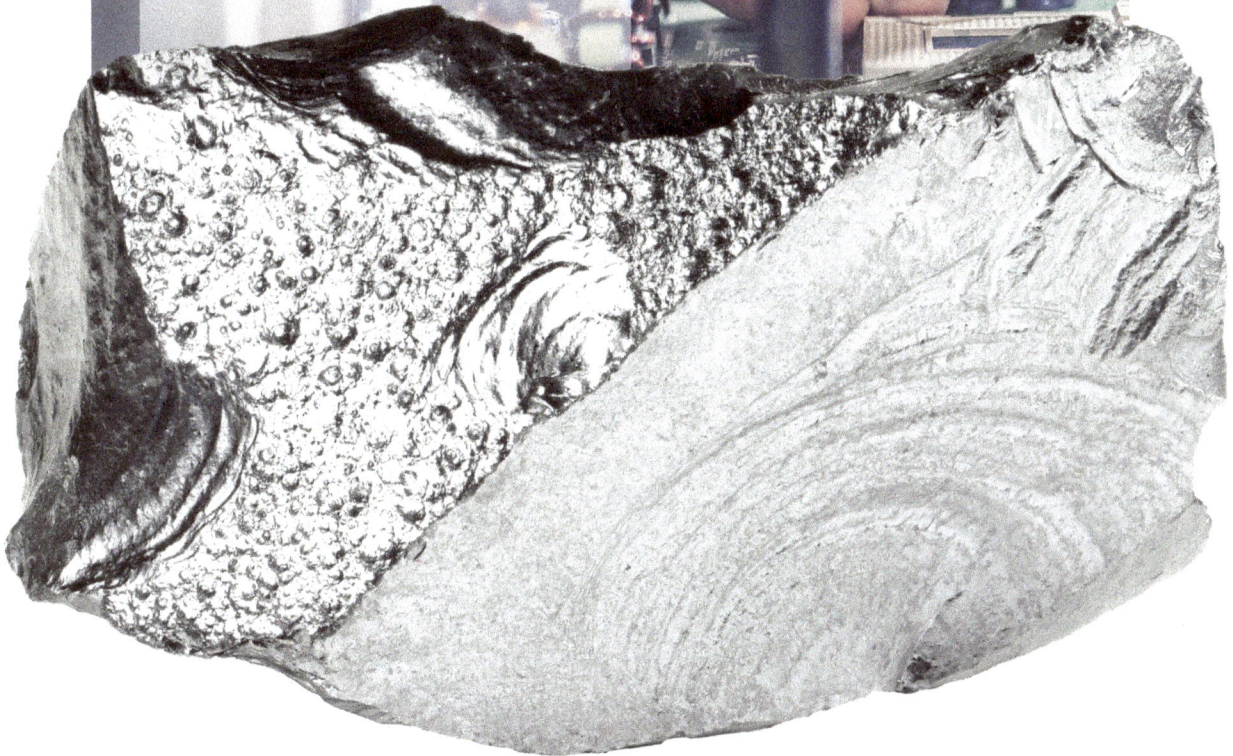

Semiconductors. Manufacturers make computer chips from a special class of materials called semiconductors. A semiconductor is a material that can conduct electricity better than an insulator but not as well as a conductor. Semiconductors allow for precise control of *electrons* (the charged particles that make up an electric current). Today, semiconductors are made from the chemical element silicon. Silicon is found in sand. It is readily available, easy to work with, and cheap.

COMPUTER CHIP HISTORY

The first electronic devices were a tangle of wires and bulky transistors. They were difficult to design and expensive to produce. Each transistor had to be connected individually, so technicians *soldered* (joined) tens of thousands of connections by hand for each computer. Some manufacturers designed sub-units with a standardized set of transistors. But this made it harder to customize a computer's design.

Silicon Valley. The area from Palo Alto southeast to San Jose is called Silicon Valley because of its many computer-related industries. The region was a hotbed for technological innovation starting in the 1900's. Stanford University, one of the foremost universities in the western United States, lies within the region. The area's western location made it a strategic hub for radar, radio, and telegraph technology. A military airfield attracted aviation and aerospace inventors.

In the 1950's, American inventor William Shockley founded Shockley Semiconductor Laboratory in Mountain View. The company was the first to produce transistors out of silicon.

Jack Kilby was an American inventor who first worked on designing transistor sub-units. But, he knew there must be a better way to solve the problems presented by ever-increasing numbers of transistors. He envisioned an entire sub-unit being "carved" out of a single block of material. Kilby began working for Texas Instruments in 1958. While most other employees were on summer vacation, he experimented with a hybrid design of transistors cast into a small metal block and connected by wires.

Robert Noyce, another American inventor, grew frustrated with Shockley's rigid management and left to co-found Fairchild Semiconductor Corporation. At Fairchild, he conceived of the *integrated circuit* (a set of transistors carved in a chip) in 1959. Noyce co-founded the company Intel in 1968 to make computer chips. Noyce earned the nickname "the mayor of Silicon Valley" for both his co-invention of the integrated circuit and his laid-back management style, which came to define many technology companies.

Intel 4004. The idea of having a single chip perform calculations did not catch on for several years. In 1969, the chip company Intel signed a contract to design and build the chips for a calculator. The calculator manufacturer proposed using eight separate chips, but to save time and resources, Intel suggested just four chips, one for each main computing process. The Italian-born **engineer** Frederico Faggin designed the 4004 chip, which performed all the calculations within the device.

MAKING COMPUTER CHIPS

Kilby's idea to "carve" a processor out of a single piece of material forms the basis for microchip manufacture today. Chips are created through the process of *photolithography*.

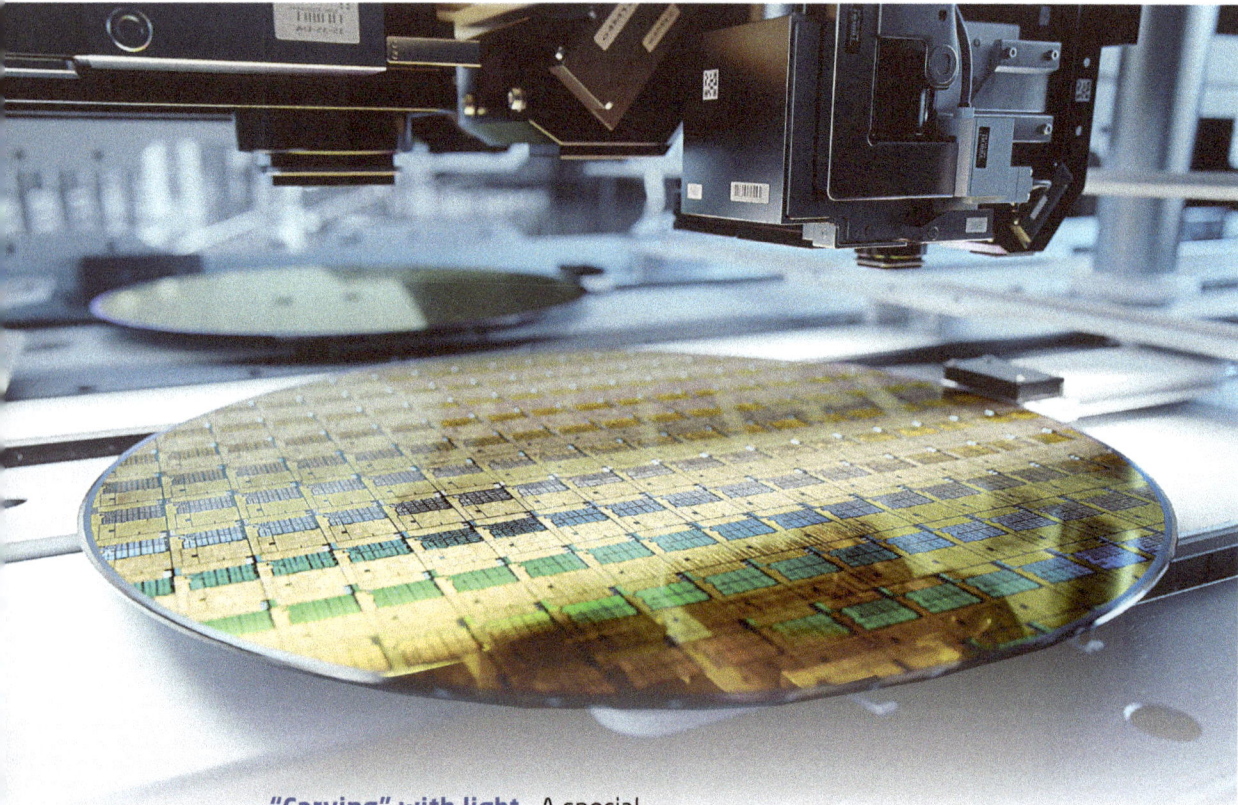

"Carving" with light. A special liquid is spread onto a wafer of silicon oxide. This liquid, called a *photoresist*, is sensitive to light and resistant to acid. The wafer is baked, hardening the photoresist. The wafer is then covered with a layer called a *mask*, which has clear sections corresponding to the transistor patterns in the chips. Powerful light shines through these sections, dissolving the photoresist beneath. The mask is removed from the wafer and acid etches the silicon oxide where the photoresist was dissolved.

Ultra-clean conditions. Any impurities that fall on a wafer during production can result in defective chips. Therefore, much of the fabrication happens in clean rooms. Such rooms have precise climate control and extensive air filtration to remove dust and other debris from the air. Workers and visitors wear special suits (nicknamed "bunny suits") when inside the clean rooms.

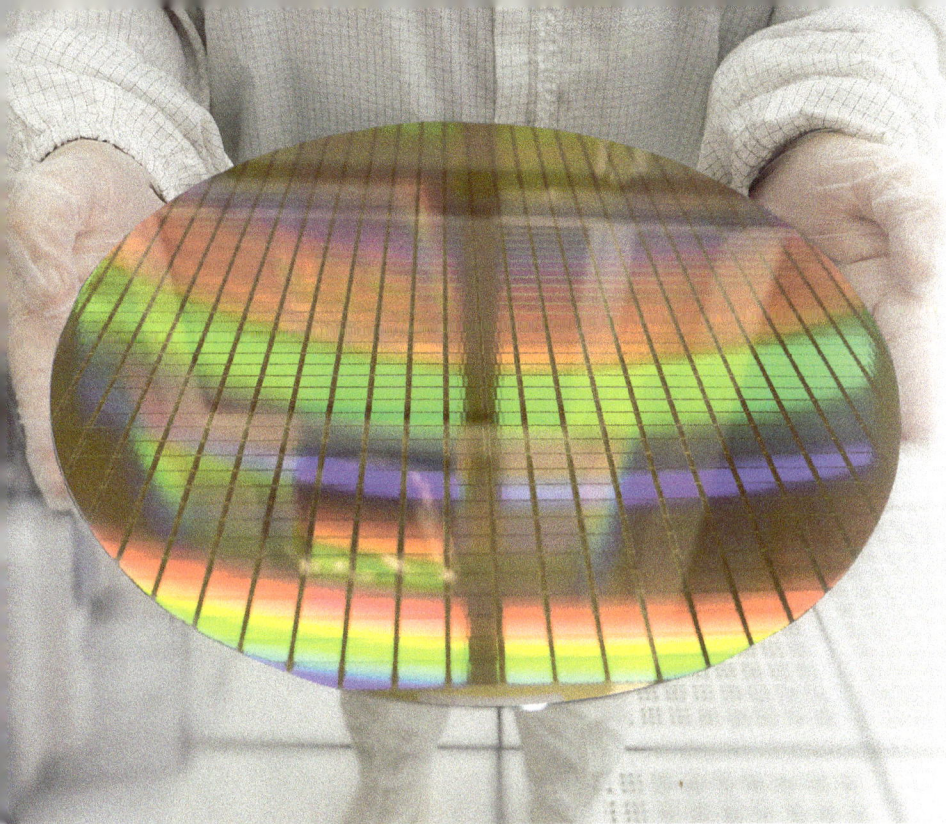

Squaring the circle. It is more efficient to etch multiple chips at the same time into large silicon wafers. The various preparations, bakes, and washings can happen faster on one large wafer than several smaller chips. A wafer is cut from a large, cylindrical *ingot* (piece) of silicon, much like a slice of lunch meat. The wafer must be circular because it is spun to evenly apply the photoresist to its surface. A square wafer would be less stable when spun and the photoresist would not be as even near the corners.

ENGINEERING CHALLENGE: OBEYING THE "LAW"

In 1965, a young technologist at Fairchild Semiconductor named Gordon Moore made an interesting observation. He noticed that, up to that time, the number of transistors per square inch that could be put on an integrated circuit had doubled every 12 months. He predicted that such a trend would continue for the next 10 years.

In 1968, Moore left Fairchild and co-founded Intel along with Robert Noyce and others. Moore's "law" became a guiding principle within the company. Intel quickly became the world's leading chip manufacturer. After Moore's initial forecast expired in 1975, he further predicted that circuit complexity would double every two years. The processing power of computer chips, compared to their cost, tended to follow the same trend.

Moore's prediction remained accurate for over 50 years. Computer chip developers etched smaller and smaller components into silicon. But this packing of transistors and increasing of power has run up against physical limits. Technologists now see the trend slowing down. It becomes ever more difficult to make transistors and circuits smaller. The very nature of matter seems to pose a limit—circuits simply cannot be made smaller than the width of a single atom.

Scientists are optimistic that new technologies will allow improvements to computer chips. But, the improvement might come at a slower rate than that currently predicted under Moore's law. It might take a breakthrough to a radically different design for the trend to continue.

The chart below plots the computing power of each new chip with the year it was released, showing how computing power has risen over time.

MOORE'S LAW

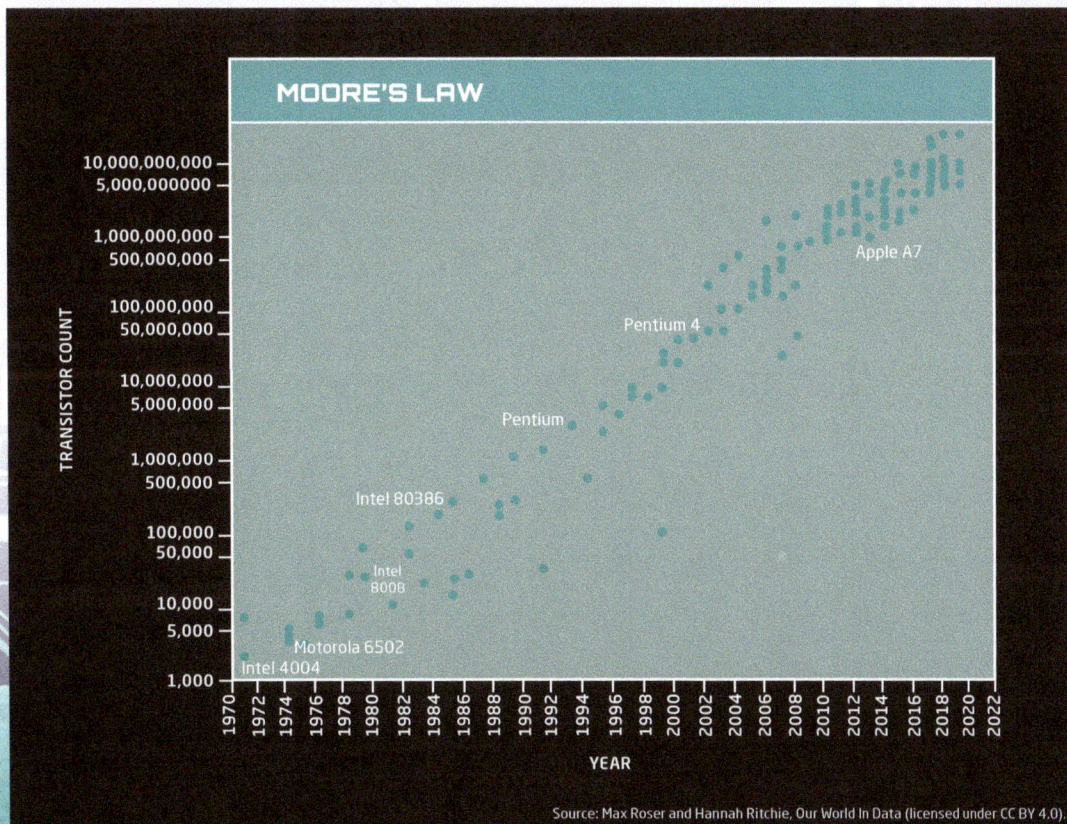

TRANSISTOR COUNT

10,000,000,000
5,000,000000

1,000,000,000
500,000,000

100,000,000
50,000,000

10,000,000
5,000,000

1,000,000
500,000

100,000
50,000

10,000
5,000

1,000

Apple A7

Pentium 4

Pentium

Intel 80386

Intel 8008

Motorola 6502
Intel 4004

YEAR

1970 1972 1974 1976 1978 1980 1982 1984 1986 1988 1990 1992 1994 1996 1998 2000 2002 2004 2006 2008 2010 2012 2014 2016 2018 2020 2022

Source: Max Roser and Hannah Ritchie, Our World In Data (licensed under CC BY 4.0).

BEYOND SILICON

Today, computer chips are two-dimensional. Electrons follow flat paths through gates and circuits. Multilevel silicon chips are not possible because the harsh process of printing the next level would damage the level beneath. But, engineers are experimenting with creating chips out of tiny structures called carbon nanotubes. Such chips could be made under less harsh conditions, allowing layers to be built up in a three-dimensional structure.

Specialized processors. The secret to speeding up processors might be to split them up again. Engineers are experimenting with separating types of computational tasks into processors specialized for those tasks. Several different processors of these kinds would be connected in a single integrated circuit. Artificial intelligence might help such a device understand how it is being used and optimize the workload sent to each processor.

Quantum computing. In a traditional computer, a transistor operates in the "on" or "off" position. But a quirk of quantum physics allows some subatomic particles to exist between states under highly specialized conditions. In a quantum computer, switches could essentially be both "on" and "off" at the same time. Quantum computers could use this strange quality to perform very powerful calculations. But they must be carefully designed to take advantage of this property—sometimes cooling the particles to extremely cold temperatures.

2 HARD DRIVES

BUILT TO REMEMBER

Computer chips can perform millions of calculations per second. But what good would a computer be if the results of its calculations disappeared immediately or were lost every time the computer was turned off? The products of all that hard work must be stored somewhere so they can be accessed in the future, whether it be the next month or the next millisecond.

A computer records results and stores information within a piece of hardware called a hard drive. Data stored on a hard drive can be read by the computer later. Importantly, a hard drive preserves the information even when you turn the computer off. When you turn on your device again later, you can thus pick up right where you left off.

INFORMATION WITHOUT WORDS

People have needed ways to keep track of things since the dawn of civilization. But paper and ink are difficult to make, and written records are fragile.

Quipus. The ancient South American Inca empire used a sturdier system of coded record-keeping: quipus. A quipu is a collection of knotted strings. Officials knotted the strings in various ways to encode information about trade goods, taxes, and other official records. Strings were also dyed various colors. Each color likely gave more information to the record keeper.

Punched cards were an important step in the development of data storage. Early computers punched results onto cards for reading and storage. Even after data storage advanced beyond punched cards, they were still used to program computers. Engineers would carefully write a program by punching holes in the cards and assembling them into a deck.

Punched cards' first job. The United States grew rapidly during the late 1800's. U.S. law requires that a census (population count) be taken every 10 years. But, the increasing complexity of the census and growing population bogged the process down. It took 7 ½ years to hand-count the 1880 census. Without changes, officials expected the *tabulation* (counting) of the 1890 census to take more than 10 years—stretching beyond the start of the next census!

The American inventor Herman Hollerith came to the rescue. Hollerith developed a system that could mechanically tabulate statistics. A census official punched a card with the information from one person and fed it into a special tabulating machine. The machine would attempt to pass a metal needle through each location where a punch could be in the card. If the needle passed through a punched hole, the needle would complete an electric circuit, adding to a counter within the machine. The 1890 census was tabulated in just 2 ½ years with this method.

STORING DATA WITH MAGNETS

Electricity and magnetism are closely linked. Many computer memory designs have taken advantage of this connection to store data. A computer can easily read from and write to magnetic storage media, so they can be reused again and again.

Drum hard drives were the first magnetic hard storage devices. Invented in the 1930's, they made use of a large, spinning drum coated with magnetized iron. Multiple fixed heads mounted around the drum could read and write information.

Magnetic tape became an industry standard for data storage beginning in the 1950's. An electronic "head" reads or writes tape unspooled from a reel. The tape is spooled onto another reel. Magnetic tape storage is slower than modern disk drives or flash drives. But, it is extremely reliable and has a high storage capacity. For these reasons, magnetic tape is still used today to back up important information.

Hard disk drive. IBM shipped the first magnetic hard disk drive in 1956. It took up an entire room and stored 5 megabytes of data—roughly the amount of data in a modern digital photograph.

Today, hard disk drives remain the workhorses of data storage. Each hard drive holds several magnetic discs. A read-write head stores and accesses data from one side of each disk.

ENGINEERING CHALLENGE: KEEPING DATA IN THE CLOUD— AND KEEPING IT COOL

Have you ever stored photos, documents, or music in the **cloud?** It's a convenient way to back up your files and access them from anywhere with an internet connection. But where do your files go? They are stored in massive **data centers!** Data centers are facilities that house computer systems. Data centers can offer a variety of computing services. Data storage is one of them.

If a computer has a single hard drive, all its data can be lost if that hard drive fails. A data center, on the other hand, stores a customer's data across multiple storage drives. It also stores more than one copy of the data. This arrangement ensures the data will never be lost, even if one or more of the data center hard drives fail.

A home user can mimic some of these advantages by linking multiple hard drives together in a desktop computer. But, the user can still lose all the computer's data if the computer is lost or damaged. Many cloud storage services store a user's data in more than one data center, so the information is not lost even if an entire data center is destroyed!

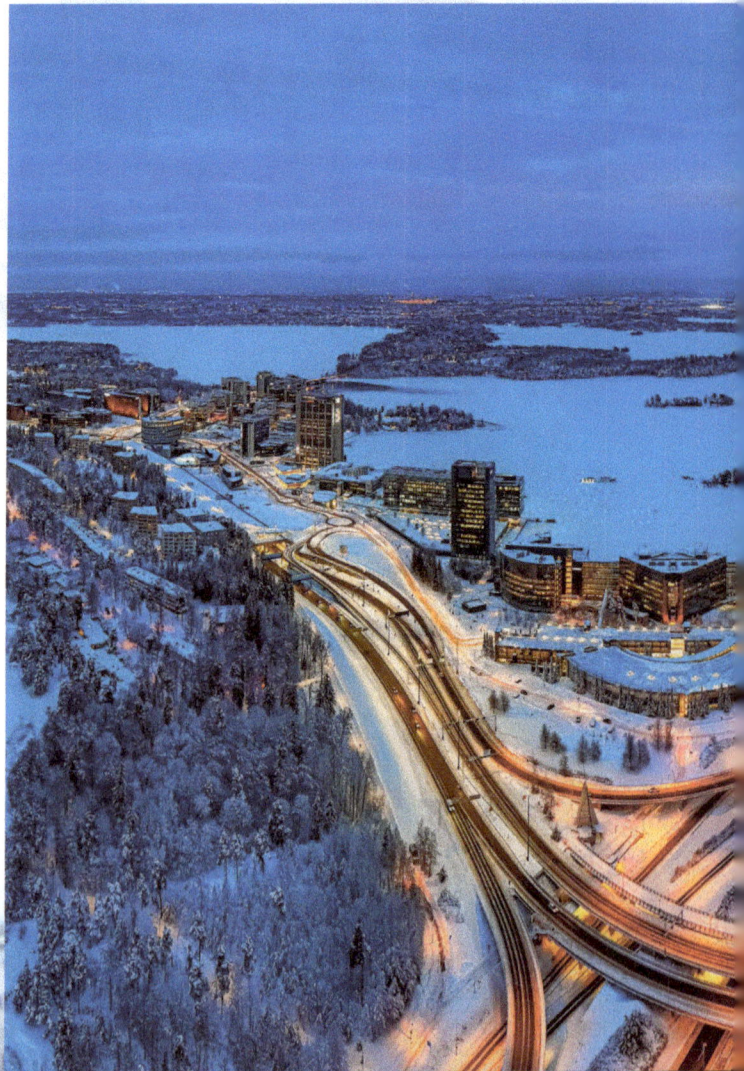

Data centers are indispensable for information storage in today's connected world. But they are not without problems. Data centers use huge amounts of electricity and emit waste heat. Computer scientists, **architects,** and engineers continue to experiment with new ways to deploy data centers to reduce their energy consumption and heat output. Some data centers have even been built in the Arctic Circle or placed at the bottom of the ocean to keep them cool! Waste heat from some data centers is captured to heat nearby homes and businesses.

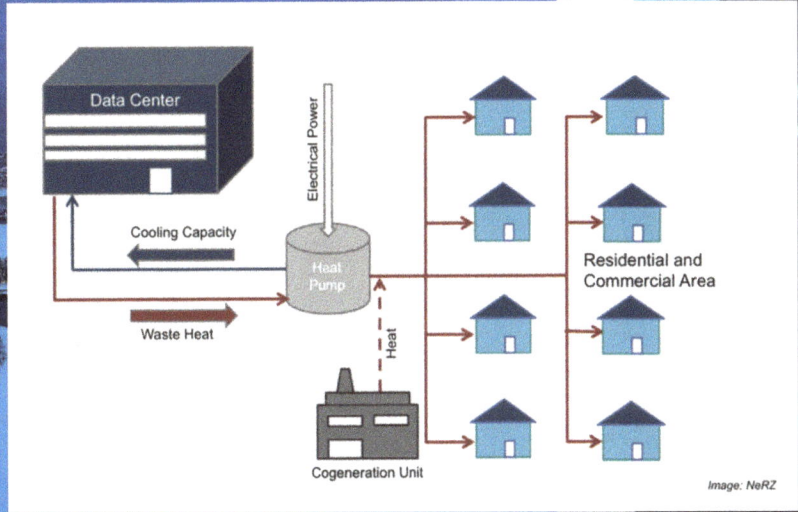

Image: NeRZ

A revolutionary new data center will run entirely on renewable energy, using its waste heat to heat nearby buildings (top). The data center is being built by the American technology company Microsoft in the northern European country of Finland (left).

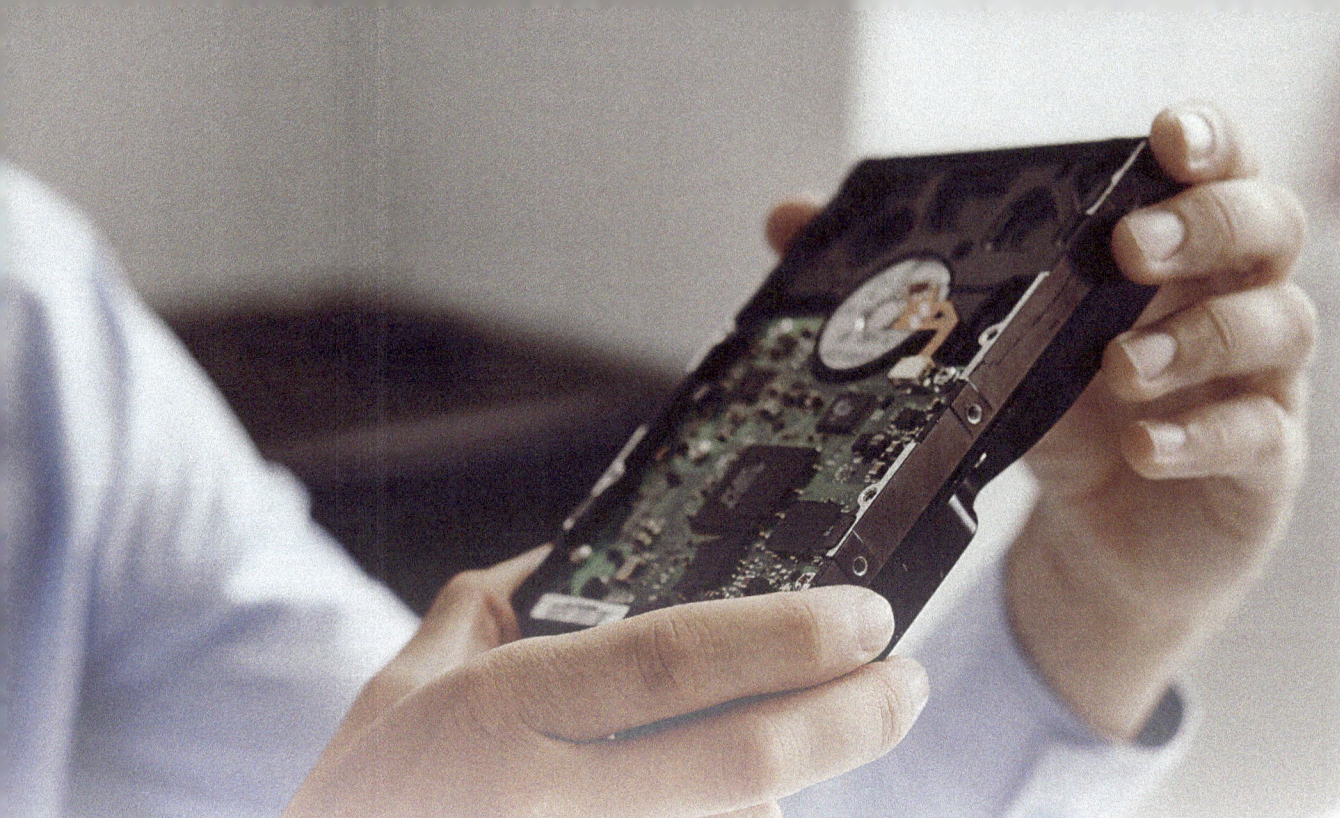

DATA STORAGE OF TODAY AND TOMORROW

Disk drives have come a long way in 50 years. No longer room-sized, hard disks fit into small, standardized bays. Today, computer enthusiasts can buy a 20-terabyte hard drive—with 4 million times more storage than the first disk drive! Engineers are developing disk drives that are quieter and use less energy.

Flash memory. Moving parts make a magnetic disk drive susceptible to failure, especially in mobile applications. But flash memory devices can store data electronically, even when turned off. With no moving parts, flash memory has enabled the development of high-capacity smartphones and made laptop computers lighter, more durable, and more energy efficient. Even desktop computers and other stationary machines have begun using solid-state hard drives.

Optical storage makes use of special discs marked with millions of microscopic pits. A **laser** detects the pits and converts their pattern into computer code. The most common optical storage media are DVD and Blu-Ray discs. The explosion of streaming services for movies, television shows, and music has reduced people's use of optical storage. But optical discs are still great for keeping permanent, high-quality copies of data.

DNA storage. Nature solved the information storage problem billions of years ago. Each cell within a living thing holds an entire copy of its **genes**—many gigabytes' worth of information—in the structure of the molecule **DNA,** or deoxyribonucleic acid. Surprisingly, we may soon be able to store data in DNA as well. The density of DNA is so great that a DNA-based server the size of a desktop computer could store all the data that humankind has ever produced.

The fifth dimension of optical storage. Optical storage may be poised for a comeback. A standard compact disc (CD) makes use of a single layer of pits engraved on a flat platter—two dimensions. A DVD has several of these layers fused together. Engineers from the University of Southampton in the United Kingdom are developing an optical disc that squeezes in an extra two "dimensions" of information, for a total of five. The extra dimensions come from the way the pits are engraved, causing them to reflect laser light in different ways. A standard-size 5D disc could hold up to 500 terabytes!

3 TOUCHSCREENS AND OTHER INPUT DEVICES

COMPUTING AT YOUR FINGERTIPS

The most powerful computer in the world would be useless if we could not tell it what to do. We give computers all kinds of instructions through *input devices.* These devices can be as high-tech as a brain-computer interface (BCI) or as simple as a keyboard. Input devices convert our movements or electrical signals into code a computer can understand.

Input devices have evolved over decades, but the humble keyboard and mouse remain standard equipment for desktop computers. Almost all mobile devices incorporate touchscreens, and touchscreens are catching on in laptops as well. A variety of other input devices are available to suit just about any taste.

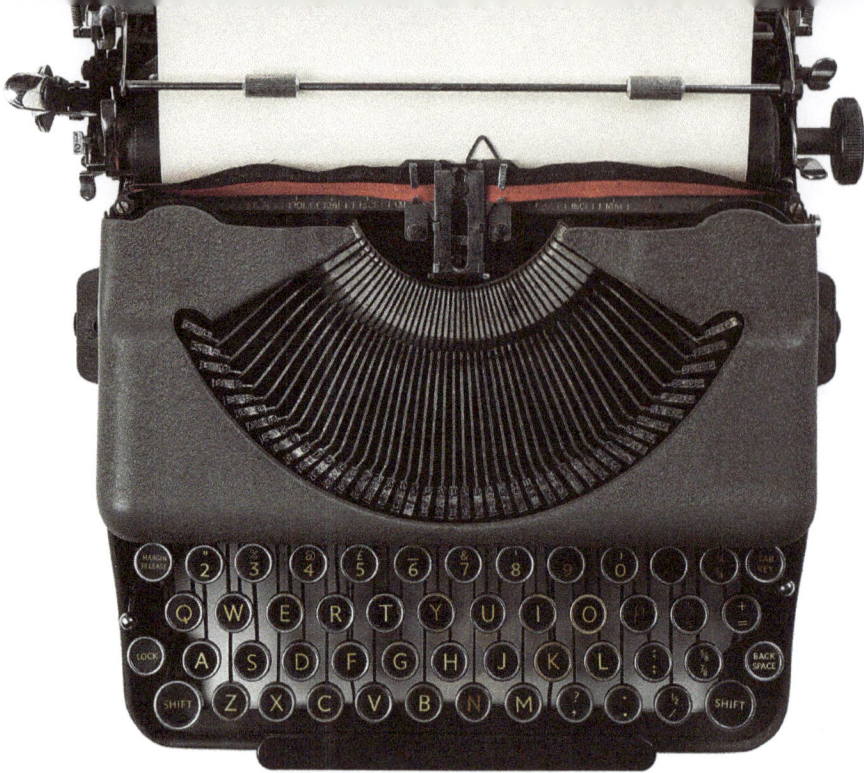

KEYBOARDS AND MICE

The modern computer keyboard descends from the mechanical typewriter. It inherited that machine's usefulness along with at least one of its quirks.

QWERTY confusion. Mechanical typewriters often jammed when certain combinations of keys were struck in rapid succession. To remedy this problem, the American inventor Christopher Latham Sholes helped developed the QWERTY layout in the 1870's. The layout, named for its first six keys, separated frequently used letter combinations to reduce jamming. Jamming is no longer a problem, but the QWERTY layout is still used in modern keyboards.

The mouse was designed in the 1960's by the American inventor Doug Englebart. The original device was a block of wood with two rolling sensor wheels attached beneath and a single button on top. During testing, the device's stout body and taillike cord earned it the nickname *the mouse*.

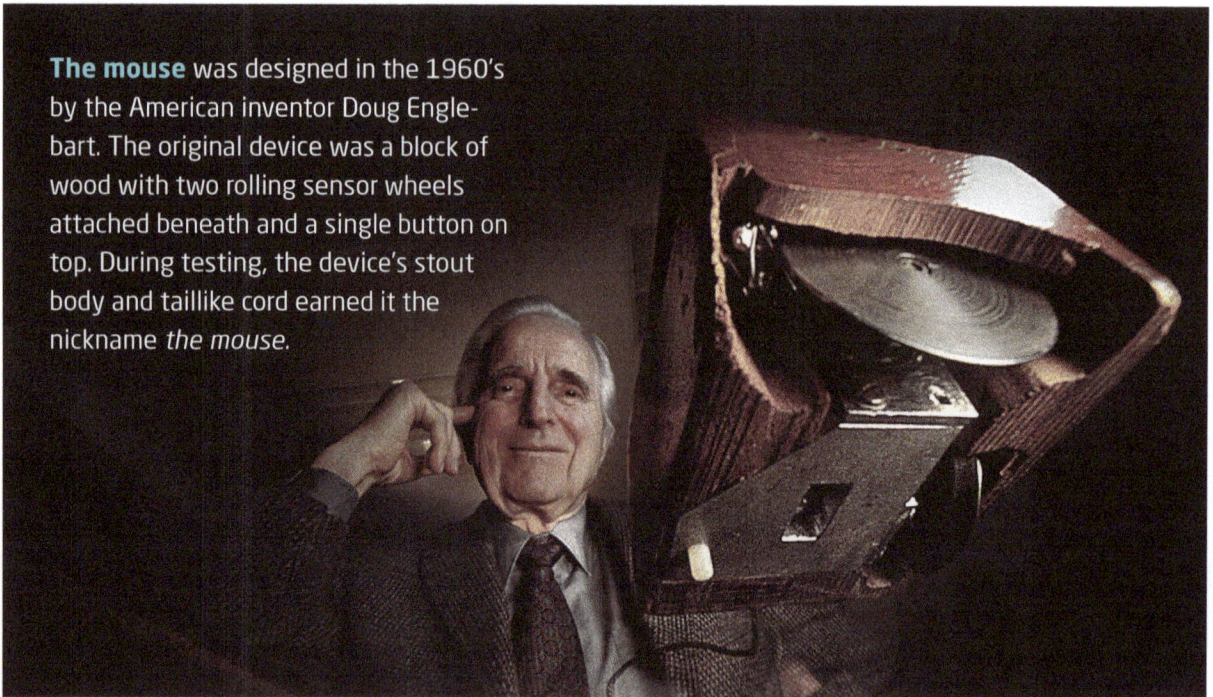

The keyset: forgotten mouse companion. If you control a mouse with your right hand, what do you do with your left? Englebart developed an input device with a few keys that enabled users to enter different commands with their left hand while their right hand used the mouse. The keyset never caught on the way the mouse did, but it attracted some die-hard users. Versions of the keyset continue to be available today.

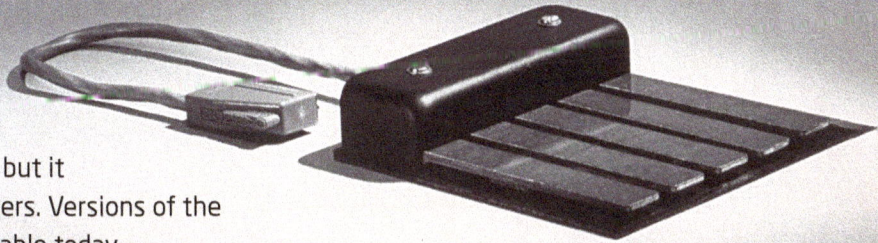

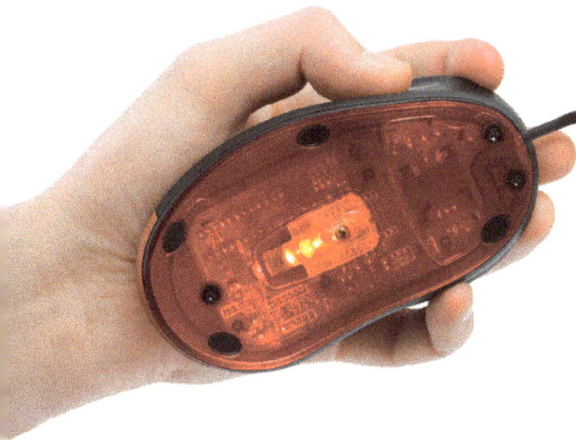

Optical mice. The traditional mechanical mouse made use of a rolling ball to track motion. But the ball could pick up dust and dirt from the rolling surface. This debris could build up on the internal components, reducing responsiveness. In 2004, the computer company Logitech released a mouse that detected movements using a small laser in its base.

KeyMouse. A company called ZZ, Inc. has added mouse sensors to a split, contoured keyboard to produce the KeyMouse. Each half of the keyboard can function as a mouse, while the user's hands remain in position on the keys. The product improves ergonomics for many users, allowing them to keep their shoulders spread and their hands on the device.

TOUCHSCREENS AND TOUCHPADS

Most touchscreens and touchpads today use capacitive technology. Capacitive screens generate a layer of electric charge. The human body conducts electricity. As a result, touching the screen with a bare finger causes a small charge to move from the screen to the finger. The charge is too weak for people to feel, but sensors in the screen register such changes.

Touchscreens invented.
The first finger-driven touchscreen was invented in 1965 by the British engineer E. A. Johnson. But it was limited in functionality and attached to a bulky cathode-ray tube (CRT) screen.

A smarter phone. Early smartphones had mechanical buttons that took up half of the device's face. Steve Jobs, the head of Apple Inc., grew dissatisfied with such devices and directed Apple to create a better smartphone. Two versions were initially developed: one with a full, capacitive touchscreen and the other with a standard screen and mechanical buttons. Jobs favored the touchscreen version and ruthlessly pushed the team to complete it.

The iPhone. Because the iPhone featured a touchscreen, more of the device's face could be dedicated to viewing. The touchscreen could also recognize more than one finger at a time, a capability called multitouch. Multitouch enabled users to control the device using intuitive gestures, such as a pinch. The iPhone dazzled consumers when it was launched in 2007 and became the model for later smartphone designs.

3 MONITORS AND TELEVISIONS

SEEING IS BELIEVING

Can you imagine using a computer or smartphone without a screen? It seems impossible. But computing did not always involve monitors. The first computers simply were not flexible or powerful enough to warrant being connected to such an expensive device. People programmed computers through punched cards and received results printed on paper.

Perhaps someday we will interact with computers primarily through **virtual reality** headsets, a set of **augmented reality** glasses, or even a direct computer-brain interface. But it's difficult to imagine that the simple screen will go away anytime soon. A screen is convenient: anyone can watch it, and its flat shape fits easily on devices or walls. Screens have come a long way since the first grainy television pictures. Further technological improvements will continue to make them even better.

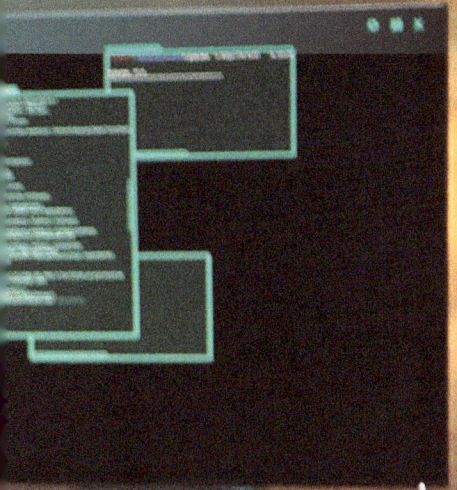

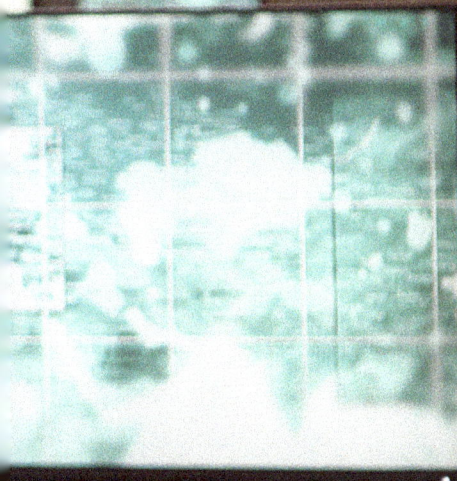

MONITOR HISTORY

The first true computer displays were **cathode-ray tubes (CRT's),** developed in the 1920's. CRT's were initially developed as memory units for computers, until engineers discovered that they had a brighter future as display products. As computers and CRT screens developed, they slowly came together.

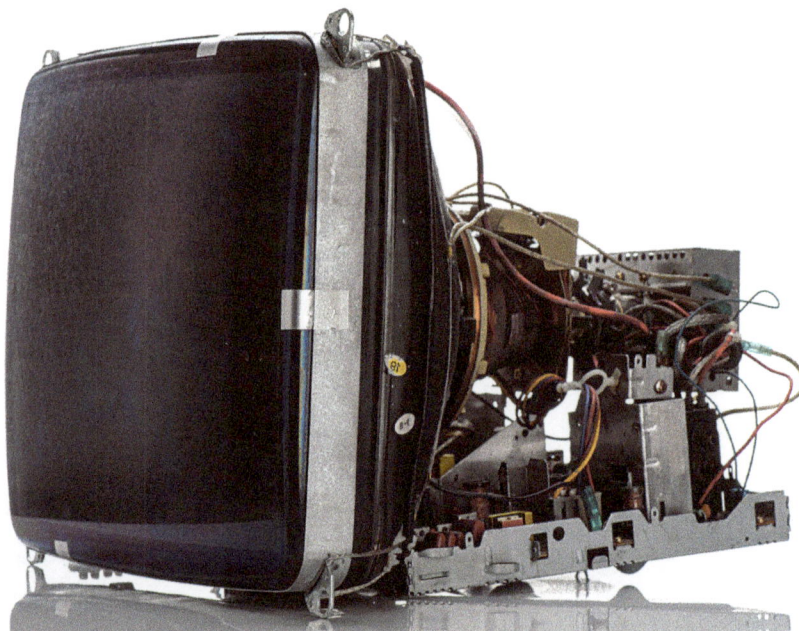

CRT headaches. CRT devices are extremely heavy. A medium-sized CRT monitor weighs around 40 pounds (20 kilograms). A medium-sized CRT television requires two strong people to lift. Cathode-ray tubes must be a certain length to display pictures. Therefore, CRT devices are quite deep and take up much desk or table space. CRT devices use lots of electricity. Cathode-ray tubes also produce a flickering image that strains the eyes of viewers.

Plasma. In 1997, the Japanese manufacturing company Fujitsu released the first commercial flat-screen television. It used a very different technology than televisions today. Inside the unit, electricity heated tiny cells of gas to **plasma** state. Plasma televisions and displays were a big step forward. But they were quite expensive and not much lighter than CRT units.

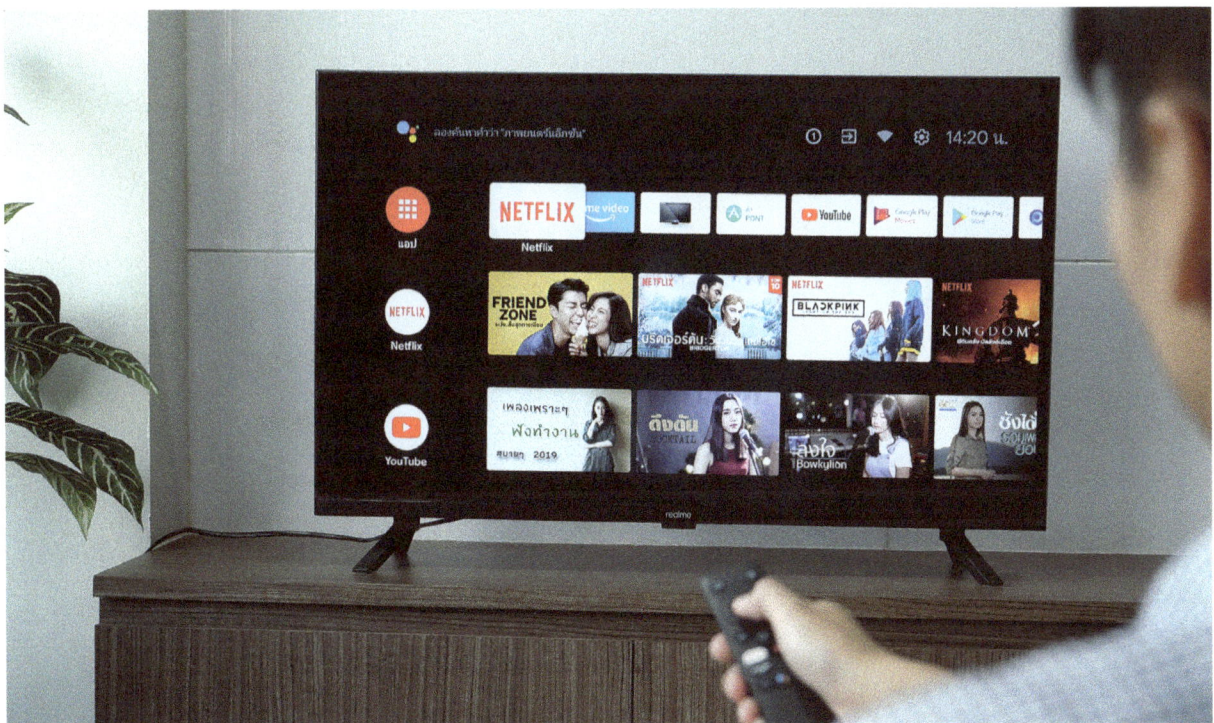

LCD and LED. In the 2000's, engineers developed monitors that use molecules called liquid crystals, which change how much light they reflect when activated by electric signals. Thousands of tiny transistors on the inner surface of a glass screen control the signals that activate the liquid crystals. Such a screen is called a liquid crystal display (LCD). At first, LCD's were illuminated by fluorescent light. But as light-emitting diode (LED) technology improved, LCD monitors switched to it, producing brighter and sharper images.

XEROX ALTO: THE FIRST PERSONAL COMPUTER

The components we are used to finding in a desktop computer all existed in the 1970's. But a computer called the Alto was first to put them all together. The Palo Alto Research Center (PARC), a division of the American office company Xerox, unveiled the Alto in 1973. They continued to improve it for the rest of the decade.

PARC created a *graphical user interface* (GUI) for the Alto. A GUI is a system that allows users to interact with a computer using icons and other visual elements displayed on a screen—rather than through typed commands. Users could point and click with a direct descendent of Doug Englebart's first mouse. Such a system required a monitor that had high **resolution** for its time. Resolution is a measure of the screen's ability to show detail.

The Alto's powerful CRT monitor supported what-you-see-is-what-you-get (WYSIWYG, pronounced *WIH zee wihg*) word processing and printing. Users could print out a page with the attached laser printer—another first—that looked exactly like what was on the screen.

PARC also equipped the Alto with a network connection. Users could compose, send, and receive messages from other Alto users. This was one of the first email systems.

Xerox failed to capitalize on the revolutionary Alto. The company finally released a production model called Star in 1981. Star was fabulously expensive, costing over $16,000 per unit (over $50,000 in 2022). Xerox marketed the computer strictly to large businesses, rather than to individuals. The company's management was not interested in the risk of refining and marketing such an expensive product for sale to consumers, who might have never even seen a computer before.

However, the Alto's legacy survived Xerox's halting efforts to commercialize it. Apple co-founder Steve Jobs visited PARC in 1979 and became fascinated with the Alto. He saw how such a simple, visual interface would allow everyone—not just specialists at large companies—to use a computer. Jobs hired away some of PARC's engineers to work on Apple's next project, the Lisa. Introduced in 1983, the Lisa was the first commercial computer to feature a mouse and a GUI.

MONITORING THE FUTURE

Monitors and television screens are not finished evolving. Engineers are developing new technology that will produce screens that are brighter, higher-resolution, and thinner than ever before, even as they use less energy.

QUANTUM LEAP

Manufacturers can add a layer of millions of tiny crystals to the backlight of an LED screen. Each crystal, called a *quantum dot,* glows at a certain color when exposed to light. The resulting display, called a quantum LED or QLED, can produce deeper color palettes than can traditional LED's.

Quantum dots

Blue LED

OLED. Backlighting can wash out darker colors and make the screen difficult to view at an angle. The next generation of flat-panel displays will make use of a layer of *organic polymers* to illuminate the picture. Organic polymers are long-chain molecules based on the element carbon. Screens that make use of them, called organic LED's or OLED's, are brighter and have greater contrast than conventional LED screens. OLED screens can also be stunningly thin—far thinner than the width of a pencil.

3D TV. In 2009, the science-fiction motion picture *Avatar* was released in 3D, triggering a rapid expansion in 3D media content. Manufacturers rolled out 3D televisions the following year. But demand for 3D movies and sales of 3D televisions quickly declined. 3D televisions were beset by competing technological standards and the need for clumsy, sometimes expensive glasses. A glasses-free 3D television was demonstrated, but users had to sit in one precise spot to experience the effect. The 3D fad fizzled as quickly as it began.

5 SUPERCOMPUTERS

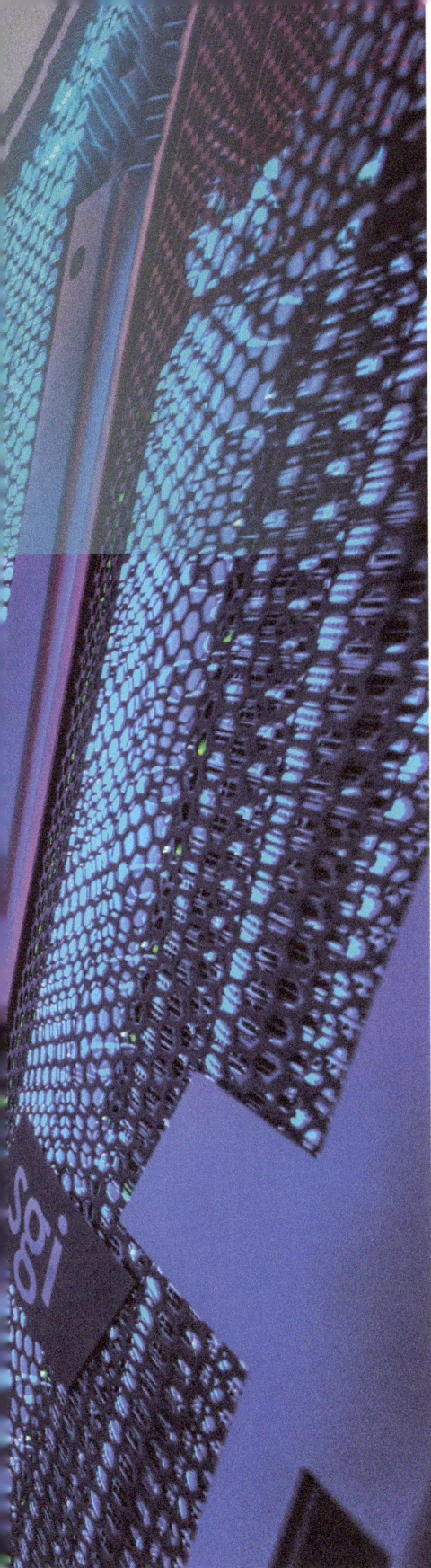

HEROES OF THE INFORMATION AGE

In the early history of computing, all computers were specialized, expensive, and rare. It took enormous machines to perform everyday calculations. Soon, computers became more powerful, affordable, and commonplace. As more scientists, engineers, and officials began using computers, they imagined ways computers could help solve extremely complex problems. The age of the supercomputer was beginning.

Supercomputers are the most powerful class of computers. Unlike personal computers, supercomputers combine thousands of processors. Several processors are grouped in a *compute node* along with dedicated memory. Each compute node works on part of the problem. The largest supercomputers can have tens of thousands of compute nodes.

The Pleiades Supercomputer

Pleiades, operated by the NASA Advanced Supercomputing (NAS) Division here at Ames, is used by scientists around the U.S. to solve complex problems for NASA missions. It is named for a star cluster in the Milky Way, which is famous for being closest to Earth and most visible to the naked eye.

The system debuted in 2008 as the third most powerful supercomputer in the world, and now has a peak processing power of 1.75 petaflops, or 1.75 quadrillion floating-point operations per second. Pleiades is 6 million times more powerful than the first NAS supercomputer, a Cray X-MP, which was installed at Ames in 1984.

Pleiades is made up of 11,776 nodes (like the one at right) with two Intel® Xeon® processors each, housed in 182 racks. It was built by the Silicon Valley-based company, SGI®.

Anatomy of a Pleiades Node

Two Intel® Xeon® processors (A) are installed at the front of the board with copper heat sinks (B) to keep them cool.

Eight dual in-line memory modules (C), or DIMM slots are used to add memory to the nodes.

The Memory Controller Hub (D) connects the processors and the rest of the hardware.

The I/O Controller Hub (E) controls all of the node's input/output and communication functions.

The Baseboard Management Controller (F) monitors sensors for changes in the node's system and environment.

The two Host Channel Adaptors (G) connect each node via InfiniBand® data cables to the rest of the supercomputer.

Seymour Cray, an American engineer, was a pioneer in the development of supercomputers. Around 1960, he designed the CDC 6600, the first device to be called a supercomputer. In 1972, Cray founded Cray Research, Inc. The company led the supercomputer field in the 1970's and 1980's. The company avoided boxy cabinets, preferring colorful, curved forms. Their machines were as stylish as they were powerful.

Supercomputer power is measured in floating-point operations per second, abbreviated FLOPS. The most powerful supercomputer is Fugaku, at the RIKEN research institute in Kobe, Japan. It is capable of over 400 petaFLOPS. A petaFLOPS is 1×10^{15} FLOPS.

What are supercomputers good for?

Simple desktop computers are quite powerful, but there are many fields in which supercomputers help advance knowledge. Supercomputers predict weather by analyzing current conditions worldwide and interpreting them through seasonal trends. They can use long-term climate data to predict the effects of global warming.

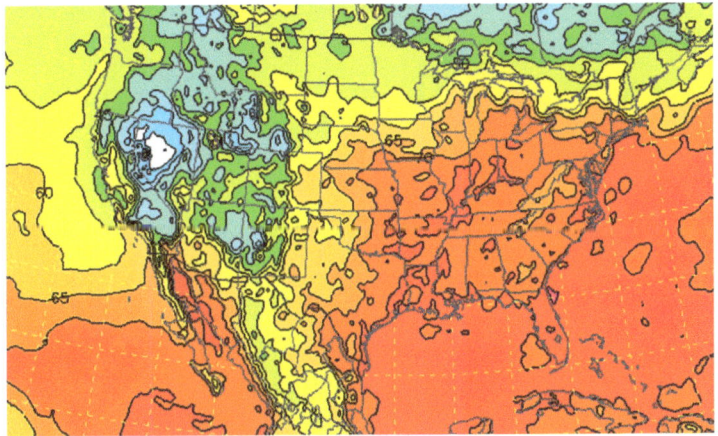

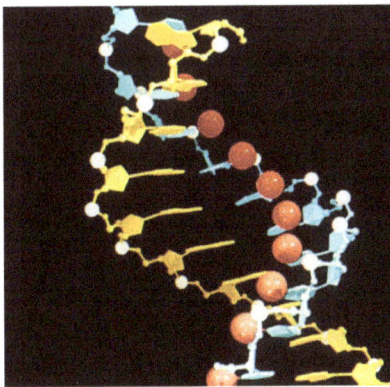

In medicine and physics. Supercomputers can also predict how atoms and molecules will react with one another, making them useful in the search for medicines. Nuclear fusion is a clean way to produce energy, but it relies on extreme conditions and expensive reactors. Supercomputers can simulate nuclear fusion reactions, allowing engineers to home in on the best reactor designs.

The race to exascale. Fugaku won't hold the crown forever. A supercomputer called Aurora at the U.S. Argonne National Laboratory is set to smash the calculation record. Aurora will be capable of over 2 exaFLOPS. An exaFLOPS is 1×10^{18} FLOPS. Even Aurora's time at the top will be limited. Other institutes are already working to best it with more powerful machines.

ENGAGE YOUR READER

Nonfiction writing often includes subject-specific vocabulary terms. Knowing the words related to the topic helps us understand the text itself.

When good readers come upon words they don't know well, they pause and try to figure them out. One tool they use is the glossary, like the one on page 4. Not every word can be defined in a glossary, though!

Authors know this, so they leave clues about words in the text. Next time you encounter a challenging word, stop and look for information about its meaning in the surrounding sentences. Sometimes authors define the term right there in the text! Other times, they'll compare the term to something you may already know. Authors even use punctuation like commas or dashes to clue you in to a word's meaning.

INSTRUCTIONS

1. Consider the list of challenge words and identify where each is used in the text. You can use the Index on page 48 to help you locate each term.

2. Explain how the author described each word. Ask yourself "what is happening in the text?" or "how is this word being used?" as you search for clues about their meanings.

3. Create your own definitions of the words. Don't just copy the dictionary definitions. Instead think about how you would tell a friend what each term means.

4. Add a visual representation for each word. Think about what you could draw that will help you remember what the words mean.

Visit www.worldbook.com/resources to download
your own graphic organizer as well as other free resources!

┌─ CHALLENGE WORDS ─────────────────────────┐

- Transistor   • Cloud

- Semiconductor   • Capacitive technology

- Processor   • Graphical user interface (GUI)

- Silicon   • Integrated circuit

└───┘

EXAMPLE

Challenge Word	Page(s)	Author's Description	Personal Definition	Visual Representation
Transistor	8-11	- tiny devices - control electric current - amplifies electric currents	Small electronic devices that control electric currents in computer chips. They can turn the current on and off and can strengthen it. Transistors make computer chips work.	
Semiconductor				

INDEX

A

Alto (computer), 38-39
Apple Inc., 33, 39
artificial intelligence, 16
augmented reality, 35
Aurora (supercomputer), 45
Avatar (movie), 41

C

capacitive technology, 32-33
carbon nanotubes, 16
cathode-ray tubes (CRT's), 32, 36
CDC 6600 (supercomputer), 44
cloud storage, 24
compute nodes, 43
computer chips, 6-17, 19
conductors, 9
Cray, Seymour, 44
Cray Research, Inc., 44

D

data centers, 24-25
data storage, 21, 22-27
desktop computers, 26, 29, 38-39
DNA storage, 27

E

electricity, 8-9, 22, 32, 36-37
electrons, 9, 16
Englebart, Doug, 30-31, 38

F

Faggin, Frederico, 11
flash memory, 26
Fugaku (supercomputer), 44-45
Fujitsu (company), 37

G

graphical user interfaces (GUI's),
    38-39

H

hard drives, 18-27
Hollerith, Herman, 21

I

IBM (company), 23
input devices, 28-33
insulators, 9
integrated circuits, 11
Intel Corporation, 11, 14-15
iPhone, 33

J

Jobs, Steve, 33, 39
Johnson, E. A., 32

K

keyboards, 30-31
Kilby, Jack, 11-12

L

laptop computers, 26, 29
lasers, 27, 31, 38
light-emitting diodes (LED's), 37,
    40-41
liquid crystal displays (LCD's), 37
Lisa (computer), 39
Logitech (company), 31

M

magnetic tape, 23
magnetism, 22-23, 26
masks (technology), 12
mice (electronic devices), 30-31,
    38-39
Microsoft Corporation, 25
monitors, 34-41
Moore, Gordon, 14-15

N

Noyce, Robert, 11, 14

O

optical mice, 31
organic LED's (OLED's), 41
organic polymers, 41

P

Palo Alto Research Center (PARC),
    38-39
photolithography, 12

photoresist, 12-13
plasma, 37
processors, 12, 16
punched cards, 21, 35

Q

quantum computers, 17
quantum dots, 40
quantum LED's (QLED's), 40
QWERTY keyboard, 30

R

resolution, 38
RIKEN (research institute), 44

S

semiconductors, 9
Shockley, William, 10
Sholes, Christopher Latham, 30
silicon, 9, 12-13
Silicon Valley, 10-11
smartphones, 26, 33
Star (computer), 39
supercomputers, 42-45

T

televisions, 35-37, 40-41
Texas Instruments (company), 11
3D media, 41
touchscreens, 28-29, 32-33
transistors, 8-11
typewriters, 30

U

U.S. Argonne National Laboratory,
    45

V

virtual reality, 35

W

wafers (technology), 12-13

X

Xerox Corporation, 38-39

Z

ZZ, Inc., 31